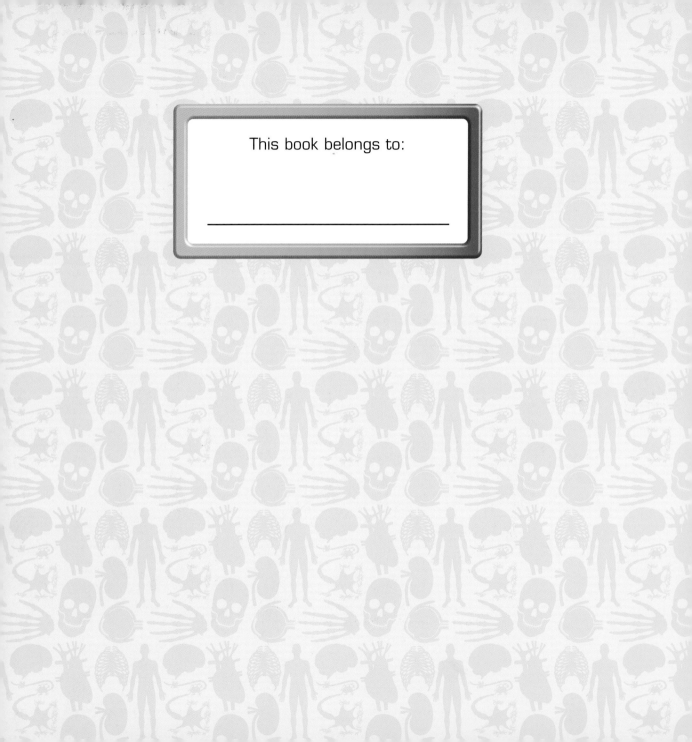

This book belongs to:

_____

Human Body

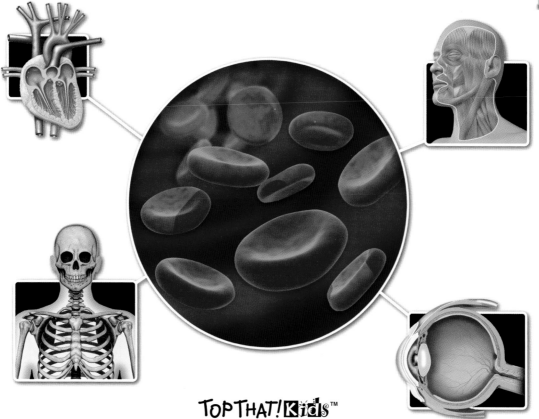

TOP THAT! Kids™

Copyright © 2008 Top That! Publishing plc
Tide Mill Way, Woodbridge, Suffolk, IP12 1AP, UK
www.topthatkids.com
Top That! Kids is a trademark
of Top That! Publishing plc

CONTENTS

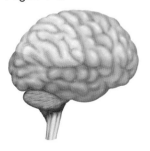
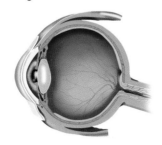
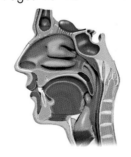

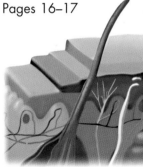

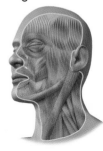

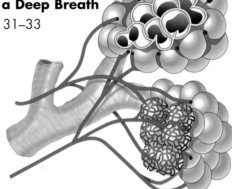

CONTENTS

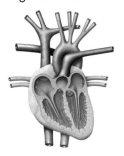

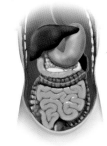
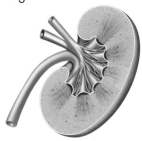
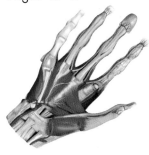

THE BRAIN IS BOSS

You should be proud to be human! That's because human brains are more complex than anything else on Earth. Together with your spinal cord, your brain controls almost every process in your body, from breathing and blinking to reading the latest Harry Potter book.

What is the brain

The brain may look like a blob of jelly, but it is one of the most extraordinary and amazing creations in the universe. It is home to a staggering one hundred billion nerve cells, or neurons! The brain is the site of your thoughts, memories and hopes. It monitors and regulates unconscious bodily processes like heart rate and breathing, and coordinates almost every movement you make.

Are human brains different to animal brains

Human brains may not be the biggest or heaviest brains in the animal kingdom, but they are certainly the most complex. Tucked inside your head is a brain that has evolved over millions of years to become the most advanced body part in the world!

Each neuron in your brain can make contact with tens of thousands of others, via tiny structures called synapses. Our brains form over one million new connections every single second! These connections are constantly changing and no two brains are alike.

FACT FILE

A newborn baby's brain grows almost three times in size during the first year!

The brain never sleeps – in fact it's always very busy. Recordings of its tiny electrical nerve signals, or 'brain waves', show that the brain stays active even during sleep, and especially when dreaming.

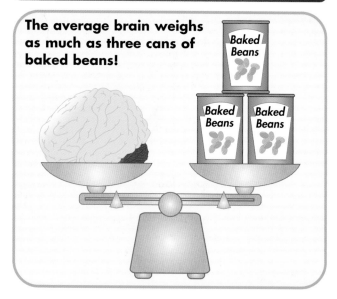

The average brain weighs as much as three cans of baked beans!

The lower part of the brain, the brain stem, deals with vital body processes we rarely think about, like breathing, heartbeat and digesting food. The smaller, wrinkled part of the lower rear brain, the cerebellum, is mainly for controlling muscles and making skilled and coordinated movements. The large upper part, the cerebrum, receives information from the eyes, ears and other senses, and carries out thinking and speaking processes.

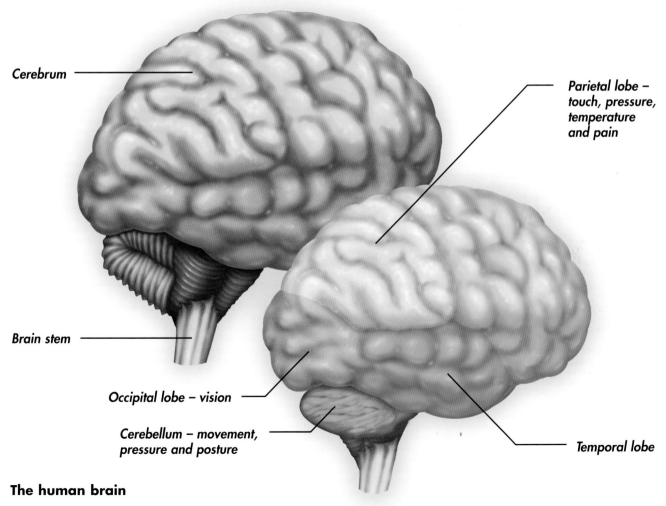

Cerebrum

Parietal lobe – touch, pressure, temperature and pain

Brain stem

Occipital lobe – vision

Cerebellum – movement, pressure and posture

Temporal lobe

The human brain

What is a CT scan ?

It is a very detailed picture of the inside of the body. A CT scanner is an advanced type of x-ray machine. It takes lots of pictures of the body and then uses a computer to put them all together, providing an image.

CT stands for 'computerised tomography'. The pictures taken by CT scans are called tomograms. They allow doctors to diagnose many serious conditions including brain tumours.

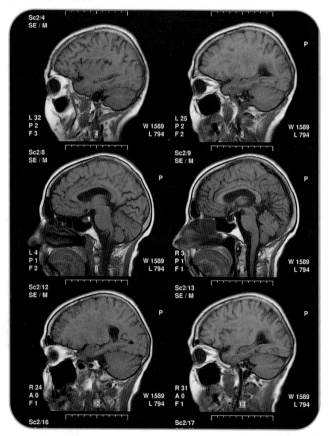

A CT scan of a female skull

Why are some people left-handed ?

No one really knows why – but we do know that the right side of the brain controls the left side of the body, including the left hand, while the brain's left side controls the body's right side.

In most people, one or the other of the brain's sides is in charge, or dominant. In about eleven people in every 100, the brain's right side is dominant so they use their left hand most.

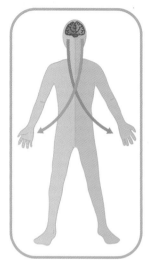

The brain hemispheres perform specific jobs

FACT FILE

Every part of the body is made up of microscopic parts called cells, and the brain has more than 500 billion of them.

The sperm whale has the largest brain in the animal kingdom. It weighs a whopping 7,800 g – over seven times more than the average human brain!

On average, a slightly higher proportion of left-handed people are likely to be creative or artistic in some way, such as artists, painters or sculptors, compared to right-handed people.

From the moment you wake up in the morning until the moment you go to sleep, your eyes are gathering information and sending it to your brain to be interpreted. Your eyes are delicate and precious. They should be shielded from sunlight and protected from harm during dangerous activities like sport.

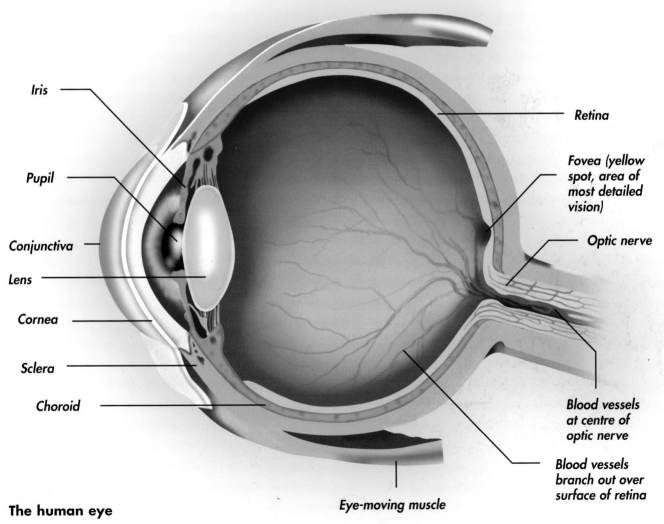

Iris

Pupil

Conjunctiva

Lens

Cornea

Sclera

Choroid

Retina

Fovea (yellow spot, area of most detailed vision)

Optic nerve

Blood vessels at centre of optic nerve

Blood vessels branch out over surface of retina

Eye-moving muscle

The human eye

7

How does the eye work ?

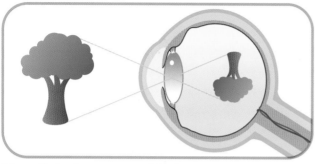

How light interacts with the eye

As the light from an object that you are looking at passes through the lens in your eye, the image becomes reversed (turned backwards) and inverted (turned upside-down). At the retina, light-sensitive nerve cells send electrical messages along the optic nerve to the brain where information about the image is processed and understood.

What does the retina do ?

The retina is located at the back of the eye. It sends information about the image we are looking at to the brain to be interpreted.

The retina contains special light sensitive cells called 'cone cells' and 'rod cells'. Rod cells process light. Cone cells process colour, and are responsible for the clarity (pureness) of an image. Cone cells are most plentiful at the back of the retina, in a place called the fovea. This is where our vision is best.

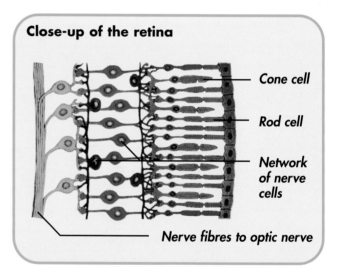

Close-up of the retina

Cone cell

Rod cell

Network of nerve cells

Nerve fibres to optic nerve

Why can't people see in the dark ?

When light fades, the cone cells of the retina (those responsible for processing colour) are no longer able to work effectively. At this time, rod cells (responsible for processing light) process any available light as shades of grey (monochrome). In this way, the human eye is less effective when it gets darker. The brain recognises the shades of grey processed by the rod cells as 'darkness'.

What do our eyelids do

Eyelids keep our eyes from drying out and prevent dirt from entering and irritating them. The rims of the human eyelid are lined with oil-producing glands. When the eye blinks, the oil from these glands is spread across the eye, ensuring that it remains moist.

Meanwhile, eyelashes serve as 'dust-catchers', catching irritants before they enter the eye.

Why do pupils change in size

The coloured ring of the eye, known as the iris, tightens and relaxes, letting the appropriate amount of light pass through the hole in the centre, called the pupil. Inside, the brightness of the light is detected by the retina, which sends messages to the brain so that it can continually adjust the iris muscles.

Why are eyes coloured

Eye colour is determined by the levels of a molecule called melanin present in the coloured tissue of the iris. The levels of melanin, in turn, are determined by genes inherited from parents. Brown is the most common eye colour in the world. Blue eyes are rarer than any other eye colour, and often found in people from Europe.

Parents' eye colour

Proportion of eye colour in children

WHAT'S THAT SMELL?

A beautiful cake fresh from the oven. Flowers during spring. A pile of dirty clothes. How do you smell these smells and thousands more? It's your nose of course! Let's be a little bit nosy and find out more about this incredible organ!

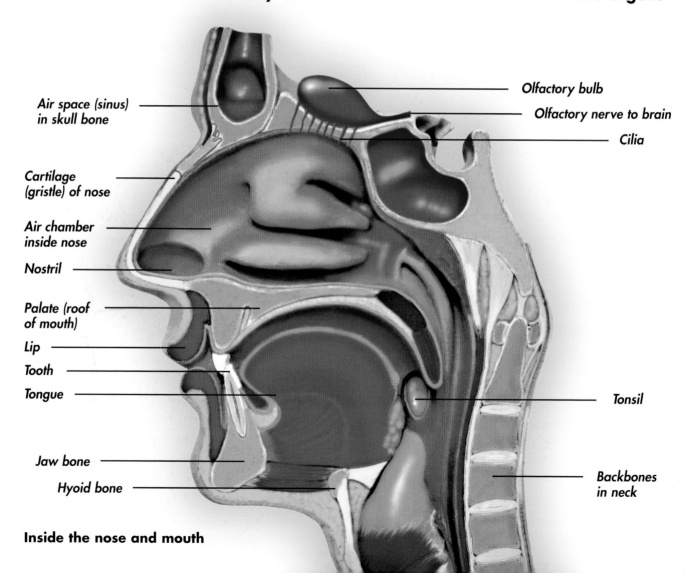

Air space (sinus) in skull bone

Cartilage (gristle) of nose

Air chamber inside nose

Nostril

Palate (roof of mouth)

Lip

Tooth

Tongue

Jaw bone

Hyoid bone

Olfactory bulb

Olfactory nerve to brain

Cilia

Tonsil

Backbones in neck

Inside the nose and mouth

How do people smell things

?

When you inhale air through your nostrils, it enters the nasal passages and travels into your nasal cavity. The nasal cavity contains special receptors, called cilia, that are sensitive to odorous molecules in the air. There are at least 10 million cilia in your nose! When they are stimulated, signals travel along the olfactory nerve to the olfactory bulb. Signals are then sent from the olfactory bulb to the brain to be interpreted as smells you may recognise, like chocolate or chips!

FACT FILE

The outer portion of the nose (the part you can see) is made from cartilage, a soft rubbery tissue that gives it its shape; it's attached to your skull at the nasal bone.

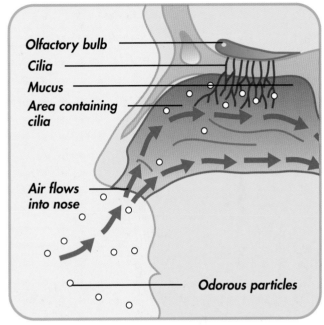

Olfactory bulb
Cilia
Mucus
Area containing cilia
Air flows into nose
Odorous particles

Odorous particles interact with the cilia

Why is smell so important

Identifying smells is your brain's way of telling you about (or helping you to make 'scents' of) your environment. Have you ever smelt smoke from a fire? Instantly, your brain interpreted the smell as a sign of danger and you knew to check where the smell was coming from. Your sense of smell can also warn you not to eat something that smells rotten, like sour milk or mouldy bread.

How is smell linked to emotion

The part of the brain that processes information from the nose is closely linked to the area of the brain that deals with emotions and memory. This partly explains why the smell of flowers often makes people feel happy, or the smell of rotten food can make people feel sick. Certain smells can remind people of special occasions. For example, the smell of mince pies may remind some people of Christmas.

You may think of the tongue when you think about taste – but you couldn't taste anything without some help from the nose! The ability to smell and taste go together because odours from foods allow us to taste more fully.

To understand how smell enhances flavour, take a bite of food and see how it tastes. Then pinch your nose and take another bite. You should notice that without smell, food tastes more bland. When you have a bad cold you may temporarily lose your sense of smell. Although your taste buds continue to work, the food you eat will taste of very little.

Hot foods give off more powerful odours than cold foods

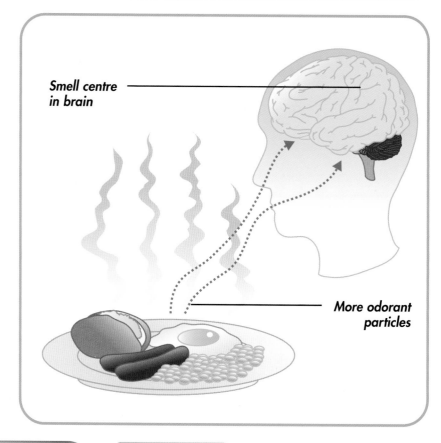

Smell centre in brain

More odorant particles

Do children have a better sense of smell ?

As people age, the cilia inside the nasal cavity become fewer. This means that young people, generally, have a far better sense of smell than older people. A similar thing happens with the taste buds in the mouth. Younger people experience smell and taste more strongly than older people do.

FACT FILE

The average person can distinguish more than 8,000 smells and odours.

Our sense of smell is 20,000 times more responsive than our sense of taste.

Your mouth is a remarkable thing. Your teeth tear up food, munching it into a wet ball. Chemicals in your saliva moisten the food, releasing its wonderful flavours. Your tongue pushes the ball of chewed food to the back of your throat, and then it is gone ...

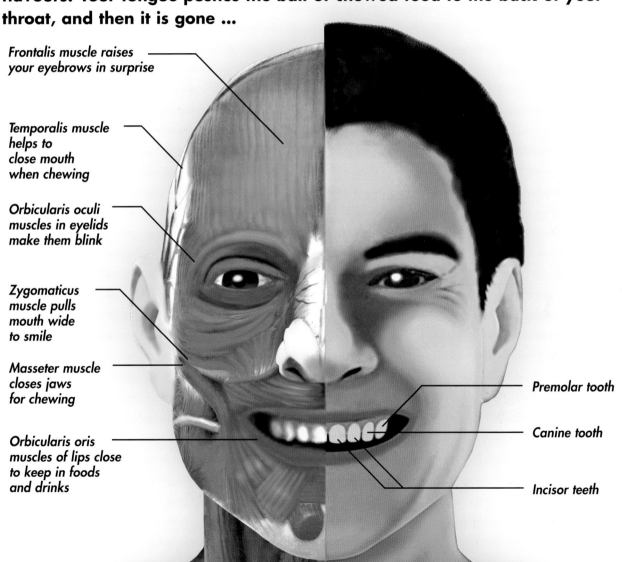

Frontalis muscle raises your eyebrows in surprise

Temporalis muscle helps to close mouth when chewing

Orbicularis oculi muscles in eyelids make them blink

Zygomaticus muscle pulls mouth wide to smile

Masseter muscle closes jaws for chewing

Orbicularis oris muscles of lips close to keep in foods and drinks

Premolar tooth

Canine tooth

Incisor teeth

What does the mouth do

The mouth is a very important part of the body. It plays a role in speech, eating, drinking, facial expressions, kissing and breathing. Bounded by the lips, the mouth links up with passages leading to the stomach and to the lungs. During eating, the tongue's main job is to push food to the teeth for chewing and to mould food into a ball for swallowing.

How is the tongue involved with taste

The tongue is covered with about 10,000 taste buds. Most taste buds are located around the sides and across the back of your tongue. They provide information about the taste of the food you eat. When your taste buds come into contact with food, they send nerve signals to the brain, where they may be interpreted as tastes you may recognise.

Why do some foods taste bitter

The taste buds on the tongue's tip sense sweet flavours best. Those along the front sides detect salty flavours, while sour flavours are detected along the rear sides. The taste buds across the back detect bitter flavourants. Therefore, we taste something bitter last of all, usually as we swallow, or just after.

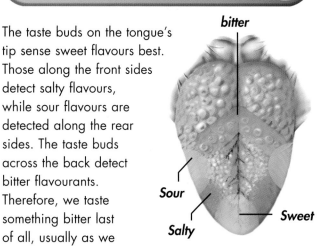

bitter

Sour

Salty

Sweet

The tongue

FACT FILE

The tongue is almost all muscle, with a thin, taste-sensitive covering. In fact it's the most flexible muscle in the body, formed from six pairs of smaller muscles all joined and working together.

The enamel covering a tooth is the hardest substance in the body. A typical tooth bites and chews over 15 million times throughout its lifetime.

Each day, six parts around the mouth (called salivary glands) make about one-and-a-half litres of saliva, or spit, to make food soft, and easy to chew and swallow.

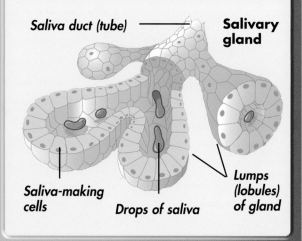

Saliva duct (tube)

Salivary gland

Saliva-making cells

Drops of saliva

Lumps (lobules) of gland

How many teeth do people have

More than 50 – but not all at once! A typical baby has 20 baby, or milk, teeth which appear from front to back, from about the age of six months to three years. By about six years, these first teeth start to fall out naturally and the 32 adult, or permanent, teeth begin to grow, again appearing from front to back. The wisdom teeth come through at 16–20 years of age – but in some people, they never develop.

How does toothpaste work

Your mouth contains over 500 different kinds of microorganisms, which eat the leftover food in your mouth. They create acid and sulphur molecules. The acid eats into teeth and causes holes, or cavities. The sulphur gives breath a foul odour.

Toothpaste rubs away these microorganisms, and removes stains caused by foods. Toothpaste also contains a substance called fluoride, which makes the teeth stronger and more resistant to acid attacks. Fluoride is the most important ingredient in toothpaste.

FACT FILE

Taste buds will only respond to liquids. Solid food in a dry mouth will produce no sensation of taste at all; it is essential for saliva to moisten solid food before it can be tasted.

The first sign of the development of teeth occurs when the foetus is only six weeks old.

Why are teeth shaped as they are

The teeth in your mouth are different sizes and shapes. This is because they perform specific jobs. At the front of the mouth, there are teeth called incisors. They are straight-edged and sharp, which enables them to bite off and slice chunks of food. Next to them is a taller, more pointed canine, which is better for tearing and pulling. Then come two premolars and three molars, which are wide with broad tops, for powerful chewing and crushing.

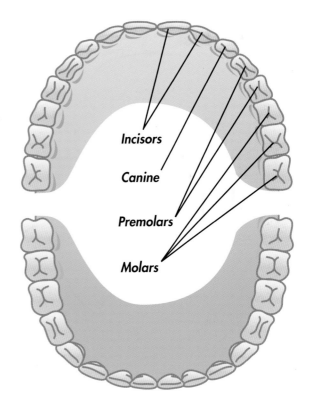

Positions of teeth in the mouth

FEELING HAIRY?

Humans are one of the hairiest creatures on Earth! Every part of your body – except for your lips, palms of your hands and soles of your feet – is covered by hair. By adulthood, you're likely to have about 5 million hairs poking out of your skin – about the same number as a gorilla!

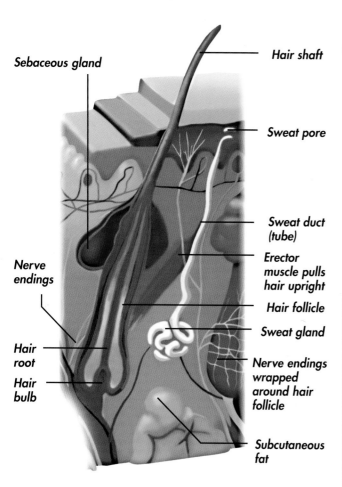

Sebaceous gland

Hair shaft

Sweat pore

Sweat duct (tube)

Erector muscle pulls hair upright

Hair follicle

Sweat gland

Nerve endings

Nerve endings wrapped around hair follicle

Hair root

Hair bulb

Subcutaneous fat

Cross section of the skin

Why do we have hair ?

Hair keeps us warm, protects the body and helps us to make sense of the world around us. When a person is cold or in danger, hair is likely to stand on end, trapping heat and making us look bigger. Hair can also help keep dust out of the eyes, ears and nose.

Sebaceous gland

Nerve endings

A hair follicle

FACT FILE

Generally, about 80 hairs are likely to fall out every day!

Nails are made of keratin, the same stuff that's found in bird feathers, antlers, cat's claws and your fingernails!

Fair hair grows more slowly than thick, dark hair.

Why does hair turn grey

Older people tend to develop grey hair because the pigment in the hair disappears and the hair turns colourless. The grey or white appearance of an individual hair is the result of the absence of something called melanin, the same stuff responsible for eye colour. A head of hair will appear grey as a result of the combination of coloured and colourless hair – not because it is actually grey.

Why are men so hairy

Male hairiness is partly to do with genes and also with the male hormone testosterone. Men can grow long, thick facial hair which forms a moustache and beard. Most women do have these hairs, and all the others that men have too, it's just that their hairs are mostly tiny and hardly visible.

How many hairs are on the head

The average human head has about 100,000 hairs. However, this number varies with age and hair colour. People with blonde hair have around 130,000 head hairs, brown haired people about 110,000 and red heads about 90,000. Older people usually have fewer hairs on their head than young adults do.

Why is some hair curly

Whether your hair is straight, wavy or curly depends on the shape of your hair shaft. A round hair shaft produces very straight hair. When the hair shaft is oval it produces wavy or curly hair. Really curly hair is produced by flattened or kidney-shaped hair shafts. The amount of moisture in the air can alter the curliness of hair, when water is forced back into the hair fibre.

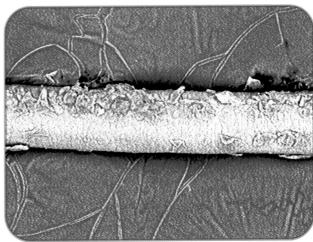

Human hair is not smooth but 'scaly' at 900 times magnification

WHAT AN EARFUL!

Not only is the ear the sense organ that detects sounds, it also plays a major role in the sense of balance and body position. The ear is a very complex organ that is divided into three parts: the outer ear, middle ear and inner ear.

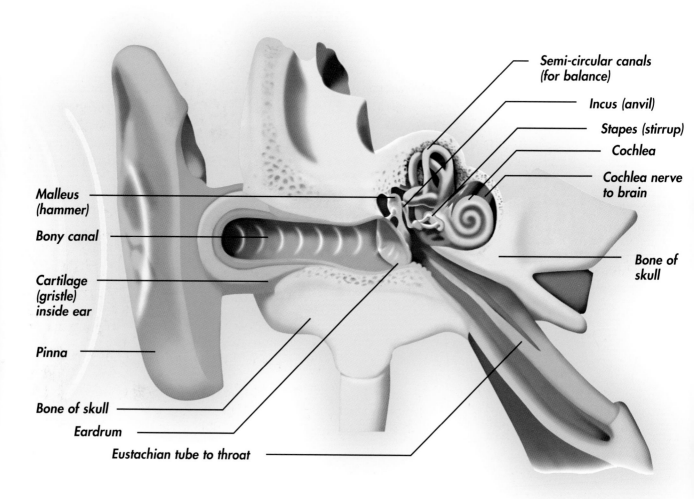

Semi-circular canals (for balance)

Incus (anvil)

Stapes (stirrup)

Cochlea

Cochlea nerve to brain

Bone of skull

Malleus (hammer)

Bony canal

Cartilage (gristle) inside ear

Pinna

Bone of skull

Eardrum

Eustachian tube to throat

The ear

How does the ear work

Sound waves travel down the ear tube, or canal, to a small sheet of thin skin called the eardrum. The eardrum vibrates as sound waves hit it. It passes its vibrations along three tiny bones, or ossicles – the malleus (hammer), incus (anvil) and stapes (stirrup) – to a part of the ear shaped like a snail and as big as a grape. This is the cochlea, in the inner ear. It is deep in the head, just behind the eye. The cochlea turns the vibrations into nerve signals and sends them along the auditory nerve to the brain.

How do ears affect balance

The ear is one of several organs responsible for maintaining the body's sense of balance. The ears, eyes, joints, skin and muscles all provide the brain with the information it needs to keep the body upright and balanced.

The parts of the ear responsible for balance are located in the inner ear – the utricle, the saccule and the semi-circular canals. When the body moves, the fluid in these tubes interacts with hair-like cells, which fire messages along nerves to the brain.

FACT FILE

The highest-pitched sounds which most people can hear are 20,000 vibrations per second – but a bat can hear sounds ten times higher.

The lowest-pitched sounds which most people can hear are 25–30 vibrations per second – but an elephant can hear sounds five times lower.

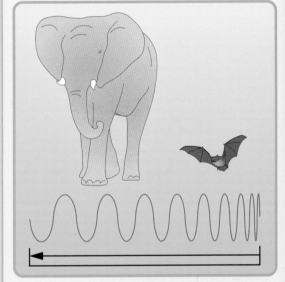

Bats can hear very high-pitched sounds. Elephants can hear very low-pitched sounds

Why do we get ear wax

The lining of the ear canal continually makes small amounts of sticky wax called cerumen. This wax plays an important role in the cleaning of the ear canal. It also helps prevent dirt, bacteria, fungi and tiny insects from reaching and harming the eardrum. Hairs grow in the ear canal for the same reason, to keep things out. Excess ear wax can press against the ear drum or block the bony canal, which causes hearing to become impaired.

What do we actually hear

We actually hear sound waves that are produced by the vibrations of air molecules. The size of these waves determines the loudness of the sound, which is measured in decibels (dB). The frequency of these waves determines the pitch of the sound, which is measured in hertz (Hz). In healthy adults, the ear is most sensitive to sounds in the middle range of 500 to 4,000 Hz.

Why can't old people hear as well as young people

As we get older, our ears become less able to detect sound waves of a higher frequency. This means that older people can't hear short, high-pitched sounds like 'ssss' and 't' as well as younger people can.

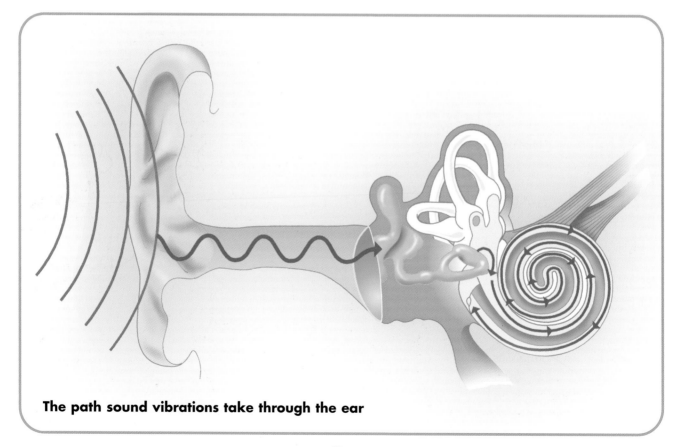

A hearing aid amplifies sound

The path sound vibrations take through the ear

THE SKELETON

Designed to provide maximum strength and mobility, the skeleton is the support network of the body, giving you shape and protecting your organs. Each bone is shaped to do a specific job and, where more flexibility is required, cartilage, ligaments and joints perform important functions.

Why have we got a skeleton ?

The skeleton provides a cage to protect the delicate parts of the body, while still allowing us to move. The ribcage, for instance, protects the lungs, heart, liver and stomach, while allowing the chest to expand and contract as we breathe.

The skeleton is also the stable frame on which the other parts of the body are hung and supported.

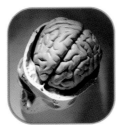

The brain

How many bones are in the human body ?

Most adults have 206 bones in their body. However, one person in 20 has an extra pair of ribs (13 pairs rather than 12). A newborn baby actually has more bones than an adult – around 350 at birth. Some of these bones fuse together as the child grows older.

The ribcage

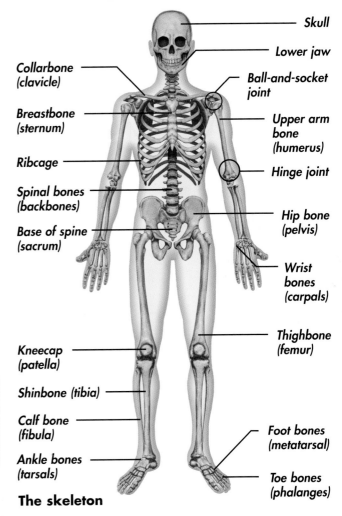

Skull

Lower jaw

Collarbone (clavicle)

Ball-and-socket joint

Breastbone (sternum)

Upper arm bone (humerus)

Ribcage

Hinge joint

Spinal bones (backbones)

Base of spine (sacrum)

Hip bone (pelvis)

Wrist bones (carpals)

Thighbone (femur)

Kneecap (patella)

Shinbone (tibia)

Calf bone (fibula)

Ankle bones (tarsals)

Foot bones (metatarsal)

Toe bones (phalanges)

The skeleton

How do joints work

When two or more bones in your body meet, they make a joint. Joints are divided into two types: mobile (synovial) and fixed (fibrous).

Synovial joints include knees, shoulders and elbows. Here, two bones are held together by ligaments, which keep them in place as the muscles make the joint move. In fibrous joints, like the joints found in your skull, are bones that are joined by tough, stringy tissue, which allows very little or no movement.

Shoulder joint

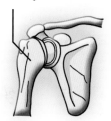

Rounded head fits in a cup-like cavity

Ball-and-socket joint (synovial joint)

Elbow joint

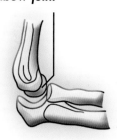

Joint moves to and fro

Hinge joint (synovial joint)

How long do broken bones take to heal

Broken bones take a varying amount of time to heal, depending on the bone broken, the age of the person and the type of break. In children, small bones take about four weeks to heal. In adults, larger bones, such as the thigh bone (femur), can take up to three months. Complicated breaks take even longer to heal.

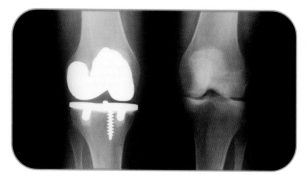

Which knee has an artificial joint?

FACT FILE

A baby is born with gaps in its skull, called fontanelles, which ensure that the skull is flexible enough to cope with birth.

A baby's skull does not form completely for up to two years. It looks like a jigsaw puzzle that joins together.

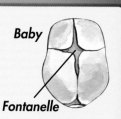

Baby

Fontanelle

Child

MUSCLE POWER

Did you know that you have more than 600 muscles in your body? They make every movement in your body possible, from the blink of an eye to a wave. You control some of your muscles and others, like your heart, do their jobs without you thinking about them at all!

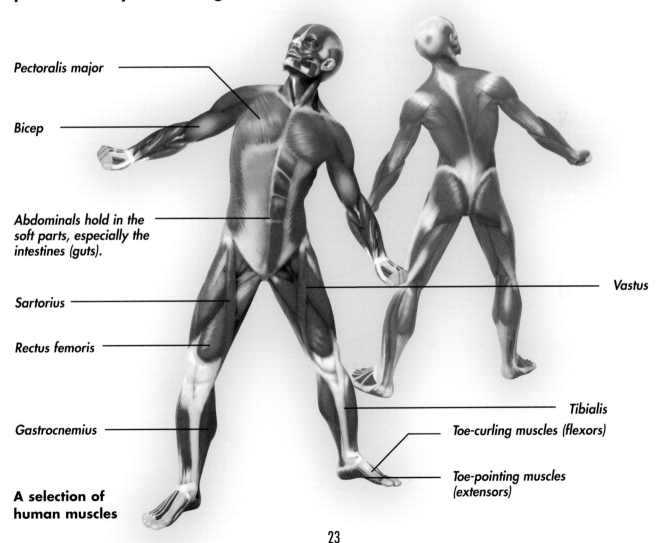

Pectoralis major

Bicep

Abdominals hold in the soft parts, especially the intestines (guts).

Sartorius

Rectus femoris

Gastrocnemius

Vastus

Tibialis

Toe-curling muscles (flexors)

Toe-pointing muscles (extensors)

A selection of human muscles

What does muscle do

Muscles move you! You have three different types of muscle in your body: skeletal muscle, smooth muscle and cardiac muscle. Skeletal muscle is attached to your bones. It enables you to consciously control most of your body, meaning that you can run, jump or walk as you choose.

Smooth muscle, however, is involuntary, which means that your brain tells it what to do without you even thinking about it. It helps to move blood through your blood vessels and food through your digestive system.

Cardiac muscle is found in the heart. This hearty muscle is responsible for pushing blood around your body.

How do muscles actually work

It all starts with orders from the brain. Special cells called neurons carry electrical messages from the brain through the nervous system to the muscles you want to move. If you're walking, your brain tells your legs to move and your arms to swing.

Skeletal muscles work in pairs. When one muscle contracts (gets shorter) the other muscle must relax (get longer). For example, your bicep must relax and lengthen when your tricep contracts and shortens. If this relationship did not exist and both muscle groups were to contract at the same time, your arm would lock and there would be no movement at all.

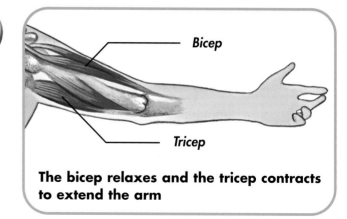

The bicep relaxes and the tricep contracts to extend the arm

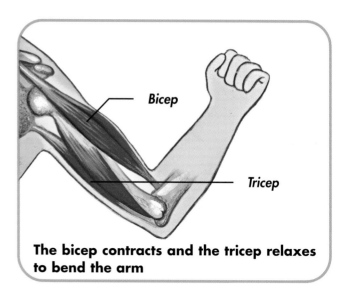

The bicep contracts and the tricep relaxes to bend the arm

FACT FILE

When you exercise, muscle fibres grow and multiply. The more you exercise, the stronger and bigger your muscles get.

What are muscles made of ?

Tens of thousand of small fibres make up each muscle. Each fibre is made up of long thin cells, called myofibril, which are packed in bundles. Each muscle has lots of these bundles – the bigger the muscle the more bundles of fibres it has.

Skeletal muscles are made up of fibres that have horizontal stripes when viewed under a microscope. Smooth muscle consists of more loosely arranged fibres. Cardiac muscle consists of fibres arranged in a neat criss-cross pattern.

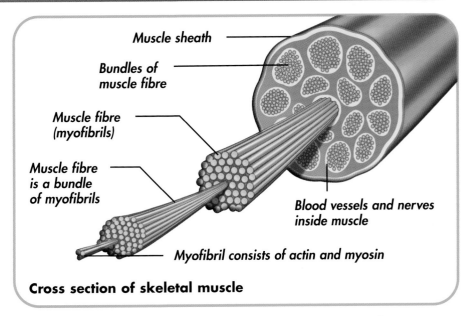

Muscle sheath

Bundles of muscle fibre

Muscle fibre (myofibrils)

Muscle fibre is a bundle of myofibrils

Blood vessels and nerves inside muscle

Myofibril consists of actin and myosin

Cross section of skeletal muscle

How many muscles does it take to smile or frown ?

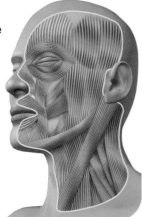

The face and head have more than 100 muscles, and we use them to make our vast range of facial expressions. It takes about 20 muscles to smile, but twice as many to frown! Scientists believe that our faces are pre-programmed to smile. A smile is used as a way of communicating with others, as a display of happiness or embarrassment.

FACT FILE

Muscle strains can happen when muscles are stretched too far. Some of the muscle fibres can be torn and there can be bruising inside the muscle.

If you exercise too long, you can get a build up of a chemical called lactic acid in your muscles, and this causes them to tighten up.

Muscles get most of their energy from glucose. Glucose is made from several types of carbohydrates such as sugar, lactose (from milk) or fructose (from fruits).

Your skin is alive! The largest organ in the human body, it's made up of many thin layers of cells in which you'll find nerves, blood vessels, hair follicles, glands and sensory receptors.

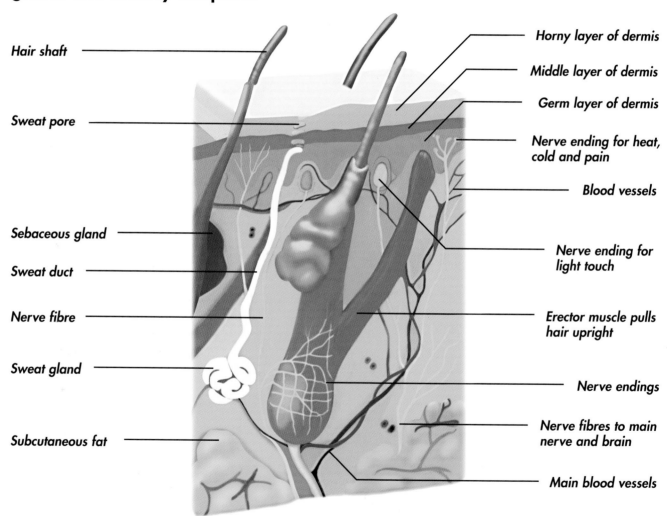

Hair shaft

Sweat pore

Sebaceous gland

Sweat duct

Nerve fibre

Sweat gland

Subcutaneous fat

Horny layer of dermis

Middle layer of dermis

Germ layer of dermis

Nerve ending for heat, cold and pain

Blood vessels

Nerve ending for light touch

Erector muscle pulls hair upright

Nerve endings

Nerve fibres to main nerve and brain

Main blood vessels

Cross section of the skin

26

Why do we have skin

Your skin doesn't just cover you. It does much more. It offers protection from the world around you, including germs and dirt. It keeps unwanted water out and keeps your body's fluids and salts in. The skin contains a variety of nerve endings that react to temperature, touch, pressure, vibration and tissue damage. It also allows the body to keep cool through sweating and softens the impact of all kinds of wear and tear.

Why do some people get sunburnt more easily

Sunburn is a burn to the skin produced by too much exposure to ultraviolet (UV light), usually from the sun. It occurs when the melanin in the skin is not able to offer enough protection against UV light. Some people have lots of melanin and others have little. People with less melanin are more likely to become sunburnt.

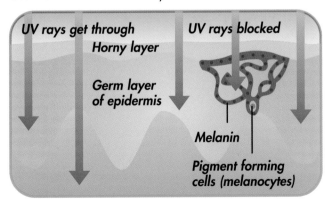

UV rays get through
Horny layer
Germ layer of epidermis
UV rays blocked
Melanin
Pigment forming cells (melanocytes)

UV rays penetrate the layers of our skin

What are goose bumps

When you are cold or scared, you sometimes develop goose bumps. Goose bumps are temporary bumps on your skin caused by the muscles that pull hairs upright. The body does this when it is cold, as it tries to insulate warm air between the hair, trapping any body heat.

Goose bumps serve a purpose in animals, fluffing up fur to make them warmer or make them look bigger and scarier.

Why do people blush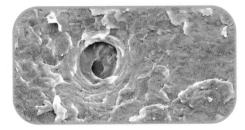

Everyone blushes. It's a normal reaction to feeling embarrassed, anxious or excited. It is caused by an increased blood flow through the tiny vessels in the skin of your cheeks. Sometimes people may not blush visibly, but may react differently, like tapping their fingers. Blushing is associated more with young people, rather than old people – and women blush more often than men do.

Close-up of a sweat pore

WHAT A NERVE!

Made up of the brain, spinal cord and billions of long nerve cells, the body's nervous system is the most complex and important network in the body. It is essential to seeing, thinking, dreaming, breathing, moving, running, sleeping, laughing, remembering, pain, pleasure, writing ... everything you do!

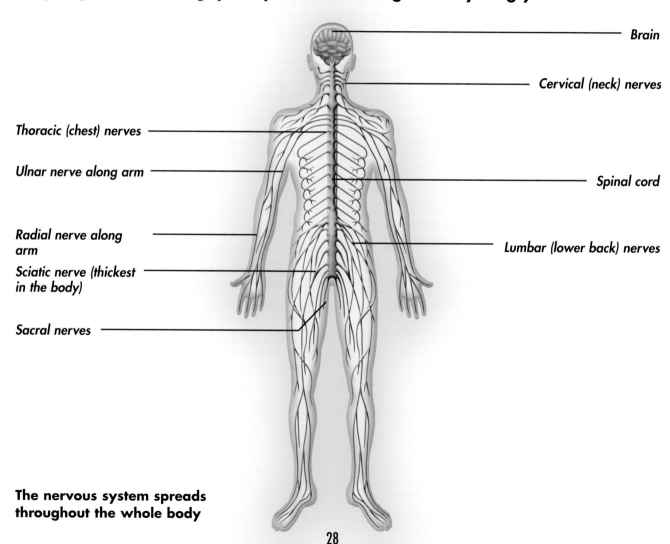

Brain

Cervical (neck) nerves

Thoracic (chest) nerves

Ulnar nerve along arm

Spinal cord

Radial nerve along arm

Lumbar (lower back) nerves

Sciatic nerve (thickest in the body)

Sacral nerves

The nervous system spreads throughout the whole body

What is the nervous system

Made up of your brain, your spinal cord and an enormous network of around 30,000 million nerves that thread throughout your entire body, the central nervous system is the control centre of your body. Your brain uses information it receives from your nerves to coordinate all of your actions and reactions. Without it, life as you know it would be impossible.

What is a nerve

Nerves, or neurons, are thin threads of cells that run throughout your entire body, carrying messages. Sensory nerves send messages to the brain. They generally connect to the brain through the spinal cord inside your spine. Motor nerves carry messages back from the brain to all the muscles in your body.

Some nerve signals travel faster than racing cars.

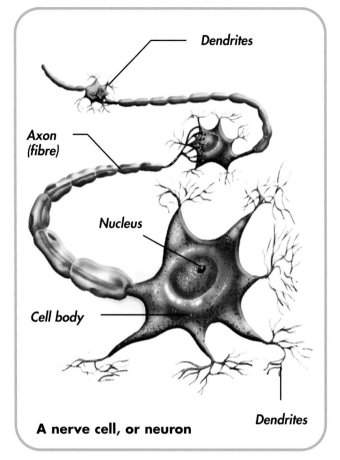

A nerve cell, or neuron

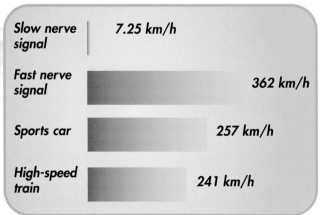

Slow nerve signal	7.25 km/h
Fast nerve signal	362 km/h
Sports car	257 km/h
High-speed train	241 km/h

FACT FILE

If you lined up all of the neurons in your body, they would stretch for an incredible 950 km!

Every second the nerves transmit nerve signals at speeds of up to 300 km/h, all around the body.

There are more nerve cells in the human brain than there are stars in the Milky Way.

What are 'pins and needles'

You may get pins and needles if a nerve becomes squashed or has its blood supply cut off. When this happens, the nerve cells can no longer send clear messages to the brain. This can result in the sensation of numbness, tingling or pricking, otherwise known as 'pins and needles'. Move, stretch and rub the affected area to get rid of the funny feeling.

What is a reflex action

Reflex actions occur quickly and automatically to help your body avoid harm. An example of a reflex action is when you touch something hot with your hand and your hand quickly moves away. The response happens quickly because it is ordered by the spinal cord rather than the brain. If the order was to come from the brain, the signals would have to travel further and your reaction would be slower. If the boy in the picture had to rely on his brain to relay reflex information to his hand, he could have hurt himself when he grabbed the hot mug.

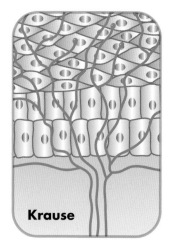

How do nerves collect information from the world around us

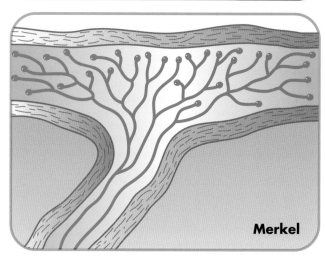

Merkel

Nerves are connected to sensory receptors in the skin. These receptors are able to collect information from the world around us. There are different nerve endings: Ruffini endings are sensitive to stretching of the skin; Merkel's discs send touch information to the brain; Krause end bulbs are sensitive to temperature.

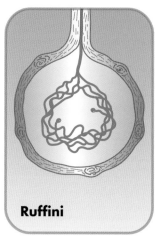

Krause

Ruffini

TAKE A DEEP BREATH

The body can survive without water for several days and without food for a week or two – but it can only survive for a few minutes without oxygen. Breathing is one of the body's most vital actions, and we do it almost without thinking – over 20,000 times each day.

Why do people breathe ?

Breathing enables the body to extract the oxygen needed for life and to rid the body of harmful carbon dioxide. Oxygen that comes into the lungs arrives at the alveoli, where it is transferred to the blood. Oxygen is used by the cells in your body to move, build, reproduce and turn food into energy. It is also at the alveoli that carbon dioxide is released from the blood and exhaled from the body.

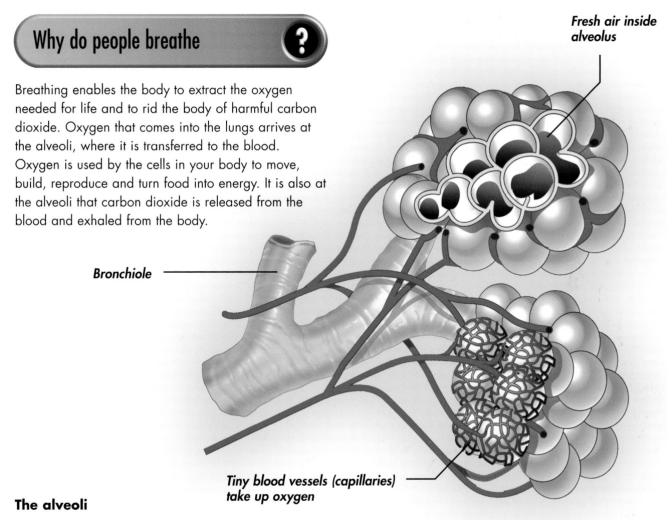

Fresh air inside alveolus

Bronchiole

Tiny blood vessels (capillaries) take up oxygen

The alveoli

Why does breathing speed up during exercise

Faster and deeper breathing supplies the body with the increased level of oxygen it requires during exercise. At the same time, when we exercise, our body produces more carbon dioxide. When we breathe faster, the heart rate increases and carbon dioxide can be moved more quickly out of the body. After carbon dioxide levels fall, our breathing returns to normal.

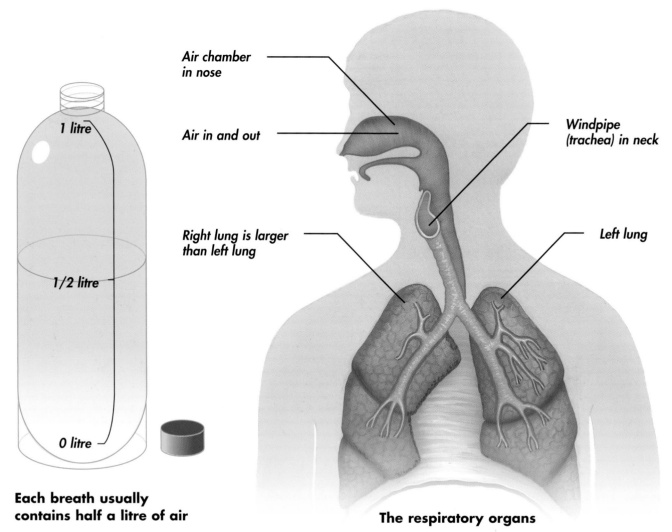

1 litre

1/2 litre

0 litre

Each breath usually contains half a litre of air

Air chamber in nose

Air in and out

Right lung is larger than left lung

Windpipe (trachea) in neck

Left lung

The respiratory organs

What causes hiccups ?

The diaphragm is a sheet of muscle in your chest, just below your ribcage. When your diaphragm tightens, your lungs expand and you breathe in air. When it loosens, your lungs contract and you breathe out.

Hiccups occur when your diaphragm suddenly jerks and air rushes over a flap (epiglottis) at the top of your windpipe, causing it to snap shut. It is the sound of this flap snapping shut that produces the 'hiccup' sound.

Why do people yawn ?

Yawning is associated with tiredness, overwork, stress or boredom. However, the specific purpose of yawning is still disputed by scientists. Some think that it cools the brain while others say that it deals with excess carbon dioxide and a lack of oxygen in the body. It could be a way of displaying tiredness or boredom to others. Some scientists think that it is a way of equalising pressure in the ears.

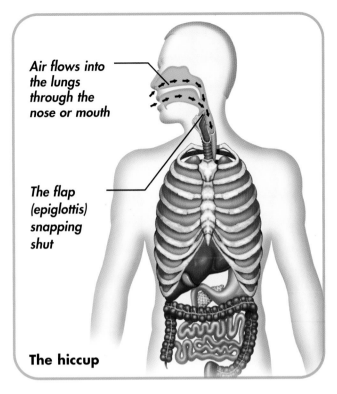

Air flows into the lungs through the nose or mouth

The flap (epiglottis) snapping shut

The hiccup

FACT FILE

On average, during a night's sleep a person breathes in and out all of the air in a typical-sized bedroom.

Plants are our perfect breathing partners! We breathe in air, use the oxygen in it, and release carbon dioxide. Plants take in carbon dioxide and release oxygen.

The tiny alveoli in both lungs, all spread out flat, would cover the area of more than 100 single beds.

HARD-WORKING HEART

The heart, the blood and the blood vessels make up the human body's cardiovascular system. The heart is responsible for pumping nutrients and gases around the body. The arteries carry blood away from the heart to the body, while the veins carry blood on its return journey back to the heart.

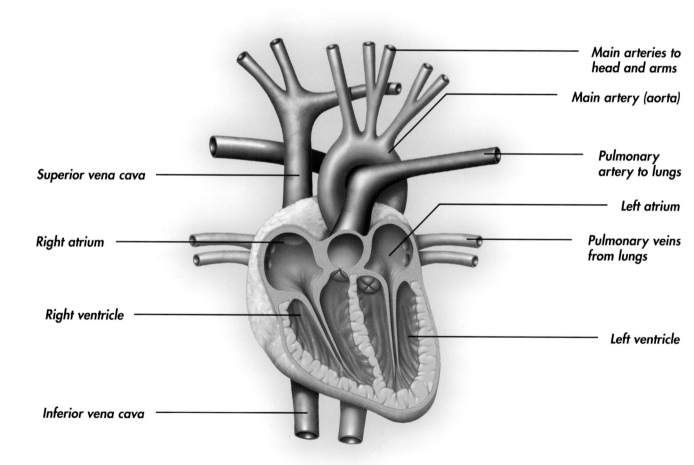

Main arteries to head and arms

Main artery (aorta)

Pulmonary artery to lungs

Left atrium

Pulmonary veins from lungs

Left ventricle

Superior vena cava

Right atrium

Right ventricle

Inferior vena cava

Cross section of the heart

What does the heart do

The heart's job is to pump blood around two separate circulations. First it pumps blood rich in oxygen from the lungs around the arteries, delivering food and oxygen to the body. Having delivered its oxygen, the blood then returns to the heart via the veins where it is pumped on its second circuit, this time up to the lungs to pick up more oxygen. The heart is also involved in communication, pumping chemical messages (hormones) around the body.

How big is the heart

The heart is about as big as the clenched fist of its owner. It weighs about 340 g, or about the same as one-and-a-half cans of baked beans. Although it is often thought of as being on the left side of the body, the heart is actually positioned more centrally in the chest, behind the lower part of the breastbone, with the larger part positioned to the left.

FACT FILE

If all the blood vessels in the body could be joined end to end, they would stretch around the world twice!

How does the heart work

Blood returns from the lungs through the pulmonary vein. It is delivered to the left atrium, which pushes the blood into the left ventricle. The left ventricle then pushes the blood through the aorta to the rest of the body. The blood returns to the heart through two large veins called the superior and inferior vena cava. It is delivered into the right atrium, which pushes it into the right ventricle. The right ventricle then sends the blood through the pulmonary artery back to the lungs.

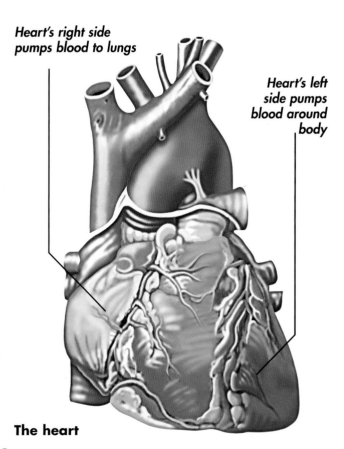

Heart's right side pumps blood to lungs

Heart's left side pumps blood around body

The heart

Does the heart ever stop beating

The heart may stop beating while the body is still alive, but only for a very short time. When you sneeze, your lungs are put under great pressure by the squeezing action of chest muscles. The heart is unable to beat properly under this pressure, so it pauses for just a moment.

Which are the thickest vessels

The thick-walled arteries and veins that carry blood under pressure to and from the heart are the thickest blood vessels in the body. At any moment, they can contain half of the body's blood. The aorta is the largest artery in the body. It carries blood rich in oxygen from the heart to the other parts of the body. Capillaries are the body's smallest blood vessels. They enable the interchange of water, oxygen, carbon dioxide, nutrients and various waste products between the blood and surrounding tissues.

What do heart valves do

Blood is pumped so quickly through the heart, you might think it would be easy for it to lose its way! But there are four small flaps, or valves, which make sure that the blood always travels in the right direction, and cannot flow back.

How often does the heart beat

Your heart beats about 30 million times a year! When you are sitting quietly, your heart beats about 60–80 times a minute. You can count how fast your heart is beating by gently feeling your pulse in your wrist. When you exercise, your heart beats much faster – about 120 times or more per minute. This sends more oxygen and energy to the hard-working muscles.

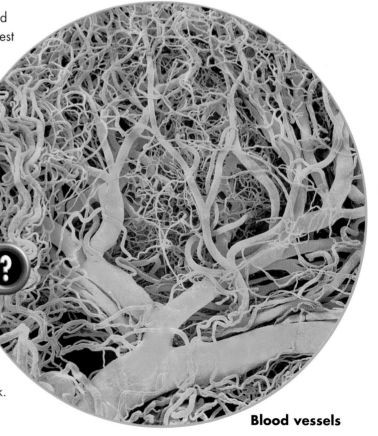

Blood vessels

RED FOR GO!

Every day the heart pumps about 10,000 litres of blood around the body – that's enough to fill 120 bath tubs! The average three-year-old has around 1 litre of blood in their body; the average adult at least five times more!

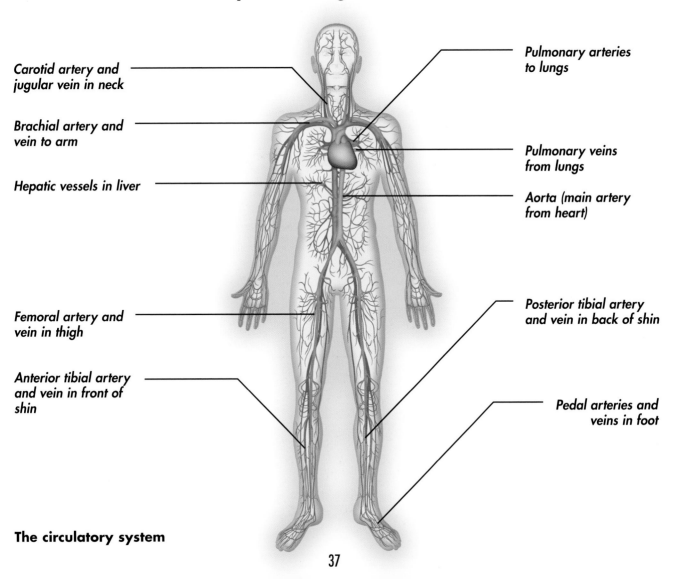

Carotid artery and jugular vein in neck

Brachial artery and vein to arm

Hepatic vessels in liver

Femoral artery and vein in thigh

Anterior tibial artery and vein in front of shin

Pulmonary arteries to lungs

Pulmonary veins from lungs

Aorta (main artery from heart)

Posterior tibial artery and vein in back of shin

Pedal arteries and veins in foot

The circulatory system

What is blood

Your blood is made up of many different cells. Most of your blood is a liquid called plasma. Red blood cells deliver oxygen to the cells in the body and carry back waste gases in exchange. White blood cells are part of your body's defence against disease. Some attack and kill germs eating them up; others attack them with chemicals. Platelets are other cells that help your body repair itself after injury.

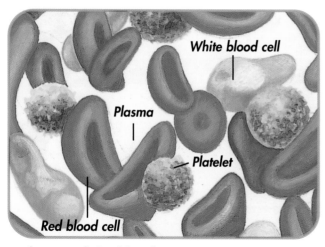

Make-up of the blood

How much blood is in the body

Blood makes up seven per cent of the body's total weight. The average adult has about five litres of blood pumping around their circulatory system. On average, men have slightly more blood than women, as women have slightly more adipose (fatty) tissues, which need very little blood supply.

Why is blood red

Because it contains red blood cells! So why are they red? They contain a red substance called haemoglobin, which carries vital oxygen from the lungs to be used all around the body. Blood that is rich in oxygen is a brighter shade of red than blood that contains little oxygen. Veins appear blue through the skin due to the interaction of light with the skin and the way that our brain interprets colour.

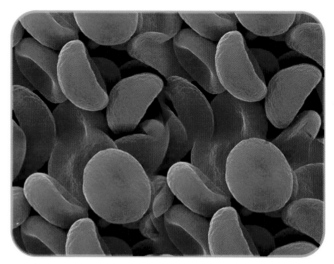

Red blood cells are doughnut shaped

Why do people give blood

Blood is given to help people who need it. People may need blood after an accident or during an operation. Before blood is given to people, it is usually split into four parts: red cells, white cells, plasma or processed plasma and platelets. Processed plasma is often given to people with haemophilia, which is a condition that stops blood from clotting properly. Haemophiliacs can lose lots of blood very quickly if they are cut, so they often require blood transfusions.

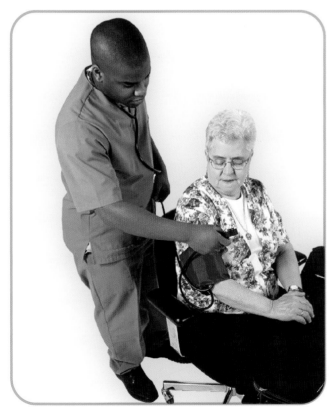

A doctor takes a patient's blood pressure

What are blood groups?

Your blood group is something that is inherited from your parents. There are 29 blood groups that are found in humans. A person's blood group is determined by the presence of substances called antigens on the surface of the red blood cells. Antigens are part of the body's natural defence mechanism. If a person needs a blood transfusion, blood of a suitable group must be chosen – the wrong type could be deadly. Most people are blood type O. Next comes A, then B, with only one person in 20 being type AB.

How does a cut heal itself

Platelets are the smallest cells in the body. They have one function only: to make blood clot when bleeding has to be stopped! Platelets are small sticky cells that are made in the bone marrow. They rush to the site of a cut, stick to the vessel wall and to each other so that a plug is formed. After this, other substances in the blood work with platelets to form a fibrin, which provides a more permanent solution to bleeding, such as a scab.

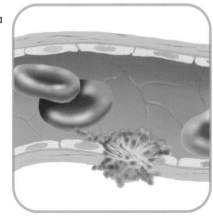

Blood clots to stop bleeding

The digestive system is made up of a variety of different organs and substances that work together to break down food into nutrients and energy that can be absorbed and used in the body's tissues and cells.

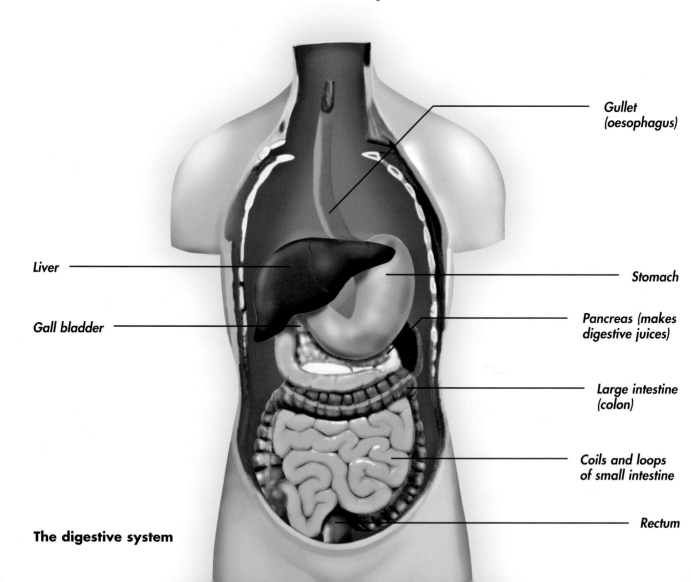

Gullet
(oesophagus)

Liver

Stomach

Gall bladder

Pancreas (makes
digestive juices)

Large intestine
(colon)

Coils and loops
of small intestine

Rectum

The digestive system

Why do people feel hungry

The body uses energy all the time, not only to move about, speak and eat, but also to make the heart beat, the lungs breathe and for cells to function. As stores of energy gradually decrease, the body signals the need for more energy by making us feel hungry.

What are the 'guts'

This is a general name for the intestines and the stomach too. The stomach is a J-shaped bag which receives swallowed food and drink. It squeezes, mashes and mixes food with powerful juices, including strong acids and digestive enzymes. From the stomach, food passes into the very long, coiled small intestine, where the digested nutrients are taken into the body.

The leftovers go on, into the shorter but wider large intestine, which takes in water and a few more nutrients. The whole process leaves faeces at the end, in the rectum, ready to be removed from the body.

FACT FILE

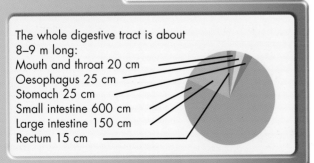

The whole digestive tract is about 8–9 m long:
Mouth and throat 20 cm
Oesophagus 25 cm
Stomach 25 cm
Small intestine 600 cm
Large intestine 150 cm
Rectum 15 cm

How does food reach the stomach

After food has been chewed in the mouth, it travels down the oesophagus and into the stomach. The oesophagus is a 25 cm long, muscular tube that contracts in waves to move swallowed food through it. The swallowing sound that can be heard during eating is the sound of the oesophagus functioning.

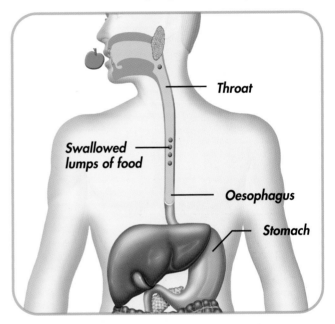

The oesophagus leads to the stomach

Why do people vomit

The muscles of the oesophagus wall are so powerful, a person could swallow while upside down. They work in reverse too, to remove bad or unwanted food from the stomach, up and out of the mouth – by vomiting!

What food takes the longest to be digested

The time for digestion depends on the type of food. Sugary and starchy foods like birthday cake break down quite quickly in the stomach and contain plenty of energy. Fruits and vegetables need more chewing and mashing. Fatty meals, especially those with lots of animals fats, take the longest to digest.

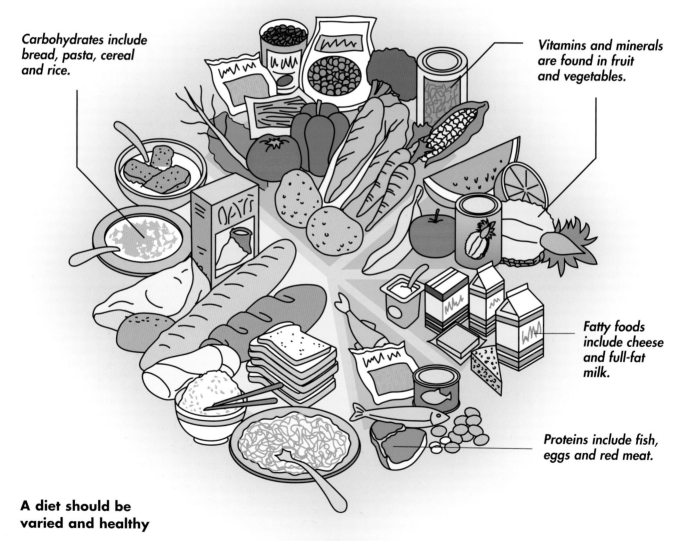

Carbohydrates include bread, pasta, cereal and rice.

Vitamins and minerals are found in fruit and vegetables.

Fatty foods include cheese and full-fat milk.

Proteins include fish, eggs and red meat.

A diet should be varied and healthy

WASTE

Excretion is the process of ridding the body of waste products that must be removed if the body is to remain healthy. The body relies on various organs for waste removal, including the kidneys, the lungs, the liver, the bladder and the intestines.

Why do people go to the toilet

We go to the toilet to rid the body of unwanted waste products. Faeces (poo!) is made up of undigested and leftover food, bits of old lining from the digestive system and old red blood cells from the liver. About 75 per cent of its weight is water, and about 10 per cent is dead gut bacteria. These bacteria live in their millions in the large intestine and help with digestion.

Produced by the kidneys, urine contains waste products produced by the body's billions of cells.

Broccoli and red meat are rich in iron and make faeces darker

FACT FILE

The kidneys receive more blood, for their size, than any other body part – about 20 per cent of all the blood pumped out from the heart.

Each day the kidneys filter the body's entire volume of blood more than 300 times.

Food drying up in the large intestine can remain there for 8 hours to two days!

Why are faeces different colours

The colour of faeces depends on the foods we eat. Lots of fatty foods make them pale. Plenty of iron-rich foods make them darker. If you don't eat enough fruits and vegetables, the faeces may be too small and hard. They may get stuck on the way out, which is known as constipation.

What do the kidneys do

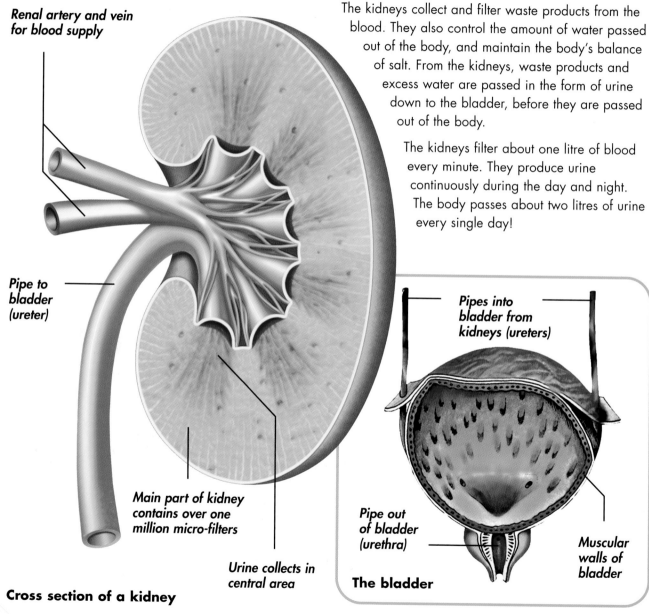

Renal artery and vein for blood supply

The kidneys collect and filter waste products from the blood. They also control the amount of water passed out of the body, and maintain the body's balance of salt. From the kidneys, waste products and excess water are passed in the form of urine down to the bladder, before they are passed out of the body.

The kidneys filter about one litre of blood every minute. They produce urine continuously during the day and night. The body passes about two litres of urine every single day!

Pipe to bladder (ureter)

Pipes into bladder from kidneys (ureters)

Main part of kidney contains over one million micro-filters

Urine collects in central area

Pipe out of bladder (urethra)

Muscular walls of bladder

The bladder

Cross section of a kidney

GET A GRIP

The hand is one of the most useful parts of the human body. From gripping a tennis racket to threading a needle, our hands are perfectly designed for performing everyday activities in the world around us.

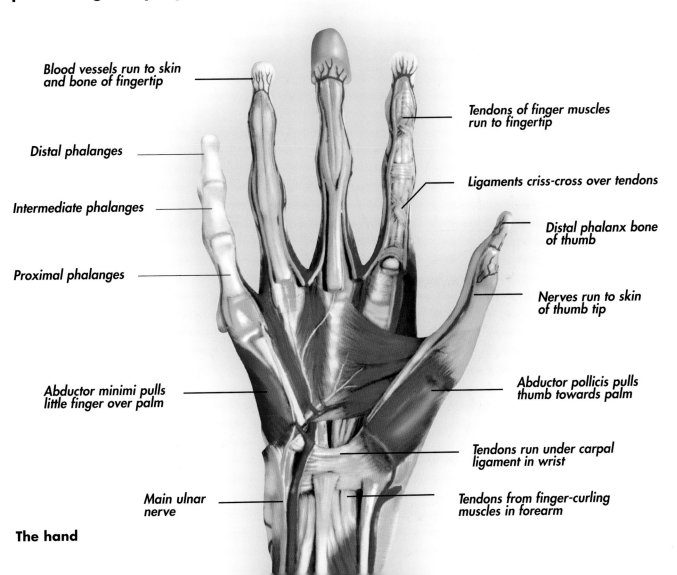

Blood vessels run to skin and bone of fingertip

Distal phalanges

Intermediate phalanges

Proximal phalanges

Abductor minimi pulls little finger over palm

Main ulnar nerve

Tendons of finger muscles run to fingertip

Ligaments criss-cross over tendons

Distal phalanx bone of thumb

Nerves run to skin of thumb tip

Abductor pollicis pulls thumb towards palm

Tendons run under carpal ligament in wrist

Tendons from finger-curling muscles in forearm

The hand

How do the hands work

The hand is truly amazing. It allows us to grip, hold, clench and perform more delicate movements like writing. It is operated by more than 50 muscles – the thumb alone has about ten muscles that move and bend it in almost every direction! Most of your hand-moving muscles are joined to bones by long, cord-like tendons that pass through the wrist. Clench your hand and you can see these tendons moving in your wrist!

How do the hands 'talk'

We communicate with our hands and fingers in many ways. For example, the forefinger held to the lips means, 'Shhhh – keep quiet'. A fist clenched by the cheek, with the thumb to the ear and the little finger to the lips, means 'I'm on the telephone.' Hands can 'talk' in another way too – by sign language. This is especially useful for people who cannot hear properly.

FACT FILE

Some people have double-jointed fingers. This means the ligaments and tendons surrounding the joints are unusually flexible, enabling them to bend or rotate them in ways other people can't.

How many bones are in the hand

The hand has 27 bones. There are five in the palm, called metacarpals, two bones in the thumb and three in each finger, all known as phalanges. There are also eight small bones, called carpals, in the wrist, which join the hand to the forearm bones.

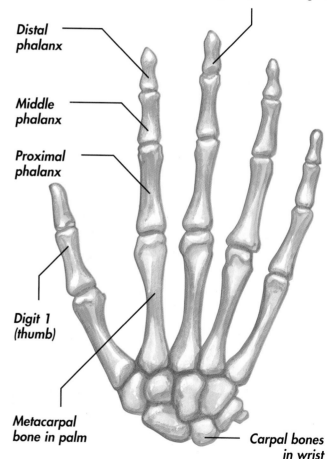

Digit 3 (middle finger)

Distal phalanx

Middle phalanx

Proximal phalanx

Digit 1 (thumb)

Metacarpal bone in palm

Carpal bones in wrist

The bones of the hand

Why is the thumb so useful

Because it can 'oppose' each finger, which means it can touch and press against each finger in turn. This enables humans to perform several different types of grip: the 'power grip', where the fingers and thumb wrap around an object; the delicate 'precision grip', which involves the tip of the thumb and your fingertips.

The power grip is used to grasp items tightly, like this tennis racket

Why do we have fingerprints

The pattern of small, curved ridges in the fingertips help fingers to grip. Smooth skin would grip much more poorly. The fingertip skin also has a thin film of sweat to improve the grip further. It's not exactly clear why everyone has different patterns, but as no print is the same, they are helpful for identifying people.

Arch

Tented arch

Pocket loop

Whorl

FACT FILE

During the Gallic wars, Julius Caesar ordered the thumbs of captured warriors to be cut off so that when they returned to their country, they would be unable to bear arms again.

BEST FOOT FORWARD

Your feet are incredibly sturdy, strong and flexible. They are able to support the weight of your entire body and still perform delicate movements like wiggling your toes. Your toenails are designed to aid balance, and the skin on the sole is the thickest of any on the human body.

How many bones are in the foot

There are 26 bones in the foot: five in the sole, two in the big toe and three in each of the other four toes. There are also seven bones in the ankle. One of these, called the calcaneus, forms the rear of the foot and the heel.

Strangely, the knobbly parts we call the ankles, on the inside and outside of the ankle joint, are not ankle bones – in fact, they are the lower ends of the long bones, called the tibia and fibula, inside the leg.

Why do men have bigger feet than women

Partly because, on average, men have larger bodies than women. However, that is not the only reason. The male hormone testosterone has an effect on bone growth and makes parts such as the nose, chin, hands and feet grow more during the teenage years in boys. No human being is perfectly symmetrical, and most of us have one foot bigger than the other.

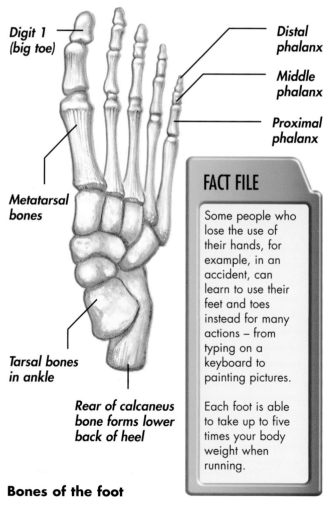

Digit 1 (big toe)

Distal phalanx

Middle phalanx

Proximal phalanx

Metatarsal bones

Tarsal bones in ankle

Rear of calcaneus bone forms lower back of heel

FACT FILE

Some people who lose the use of their hands, for example, in an accident, can learn to use their feet and toes instead for many actions – from typing on a keyboard to painting pictures.

Each foot is able to take up to five times your body weight when running.

Bones of the foot

How much pressure does the foot have to absorb?

With every step you take, your feet have to absorb a pressure that is about one and a half times the weight of your body. If you run or jump, the force may be many times your body weight. You can see how the pressure varies during slow walking, jogging and leaping; just look at the depth of the different parts of a footprint in damp mud or sand.

Walking **Jogging** **Running**

Running will leave the deepest footprints

Why do feet have toenails

Some scientists think that toenails aid walking, by making the toes stronger and stiffer for a better grip

on the ground. Long ago, when our distant ape-like ancestors lived in trees, their toes were as good at holding and gripping branches as their hands. Toenails may also be involved in balance and coordination.

Nails helped our ancestors grip more easily with their toes

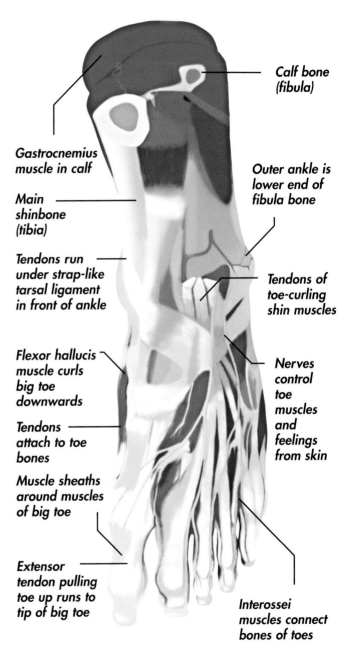

Calf bone (fibula)

Gastrocnemius muscle in calf

Main shinbone (tibia)

Outer ankle is lower end of fibula bone

Tendons run under strap-like tarsal ligament in front of ankle

Tendons of toe-curling shin muscles

Flexor hallucis muscle curls big toe downwards

Nerves control toe muscles and feelings from skin

Tendons attach to toe bones

Muscle sheaths around muscles of big toe

Extensor tendon pulling toe up runs to tip of big toe

Interossei muscles connect bones of toes

The muscles of the foot and ankle

EXPECTING A BABY

Everyone begins life in the same way – as a tiny fertilised egg, smaller than the full stop at the end of this sentence. The development of a baby in the mother's womb is one of the most amazing parts of the human life cycle.

How long does a baby spend in the womb ?

Even though a baby is thought of as 'nought' or 'zero' when it is born, at birth it is already nine months old – this is the length of time spent developing in the mother's womb during pregnancy.

How does life begin ?

Life begins with a fertilised egg, or embryo. This involves the coming together of two cells. One is the egg cell, produced inside the mother's sex organs, or ovaries. Each month one of them becomes larger and ripe, and an egg is released into the Fallopian tube (oviduct). Here it may meet the other cell, a sperm cell, which comes from the father. Sperm are made in the male sex organs, or testes. When egg and sperm join, it is called fertilisation.

FACT FILE

Throughout a woman's lifetime, she will produce 400–500 egg cells.

The human heart starts to beat 21 days after conception.

From fertilised egg to full-grown foetus, body weight increases six billion (6,000,000,000) times!

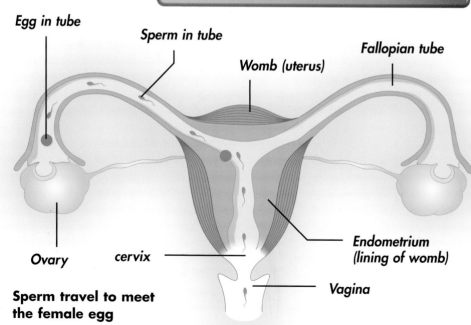

Egg in tube

Sperm in tube

Womb (uterus)

Fallopian tube

Ovary

cervix

Endometrium (lining of womb)

Vagina

Sperm travel to meet the female egg

What is an embryo ?

The embryo stage lasts from fertilisation to eight weeks later. The fertilised egg moves along the Fallopian tube into the womb, where it sinks into the blood-rich lining. All the time the embryo is dividing and re-dividing to form a small group of cells called the morula. By eight weeks this group of cells has developed to form the main parts of the human body, yet the curled-up embryo is hardly larger than a grape.

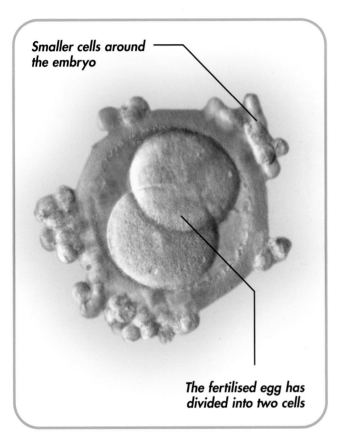

Smaller cells around the embryo

The fertilised egg has divided into two cells

A two-cell embryo

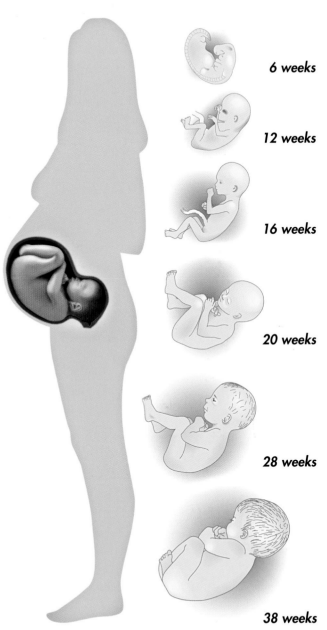

6 weeks

12 weeks

16 weeks

20 weeks

28 weeks

38 weeks

Stages of growth

51

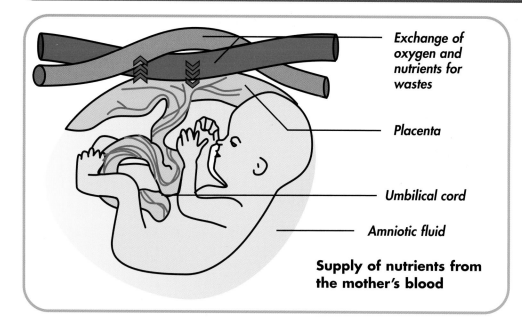

Exchange of oxygen and nutrients for wastes

Placenta

Umbilical cord

Amniotic fluid

Supply of nutrients from the mother's blood

After eight weeks, the embryo is called a foetus. It continues to grow fast as smaller details of its body form, like eyelids and fingernails. By the time of birth (depending on the parents' size and ethnic group) the baby weighs about 3–3.5 kg.

How does a baby breathe in the womb

It can't – it is surrounded by fluid in the womb! The foetus receives all the oxygen it needs from its mother. This vital exchange is carried out by the placenta, to which the foetus is attached by the umbilical cord. At the placenta, the foetus's blood vessels pass closely to the mother's blood vessels and the exchange of oxygen, food and waste products takes place. The placenta also acts as a barrier to protect the foetus from harmful substances.

Umbilical arteries carry deoxygenated blood from the baby; the umbilical vein carries oxygen to the baby

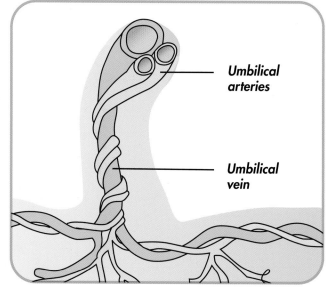

Umbilical arteries

Umbilical vein

A NEW LIFE BEGINS

A baby spends about 38–40 weeks growing inside its mother's womb (uterus), where it is warm, well fed and protected. The lights, sounds, actions and fresh air of birth come as a big shock!

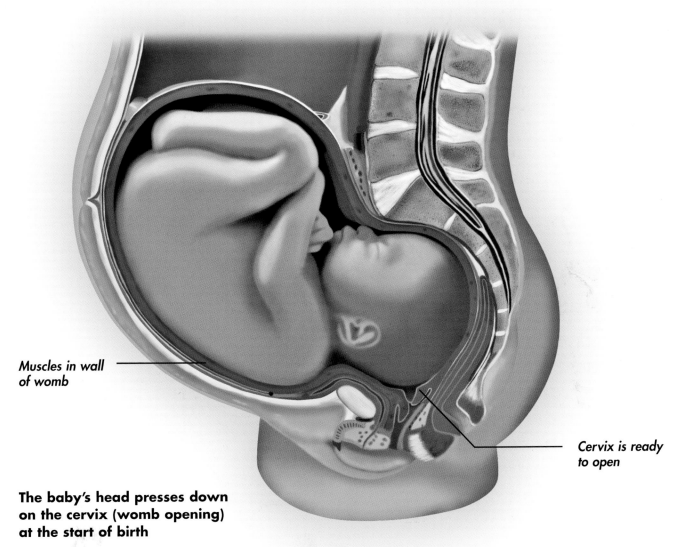

Muscles in wall of womb

Cervix is ready to open

The baby's head presses down on the cervix (womb opening) at the start of birth

How is a baby born

The womb in which the baby grows has powerful muscles in its walls, and a narrow, pocket-like opening called the cervix. At birth, the womb muscles shorten or contract powerfully, and the cervix becomes much wider. The muscles push the baby through the dilated cervix, along a tunnel called the vagina, or birth canal, and out of the mother's body.

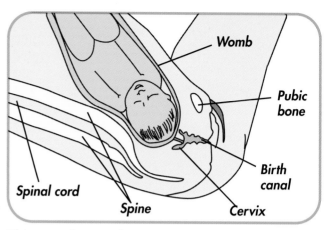

The womb muscles contract

How long does birth take

On average, it takes 14–17 hours for a mother to deliver her first baby, much shorter for her second baby, even shorter for the third, and so on. The longest part of birth is the first stage, when the womb muscles start to contract. These muscle contractions can last for a number of days. The second stage, delivery, can last from anywhere between 15 minutes and two hours. The placenta is usually delivered between 15 minutes and one hour after the baby's birth.

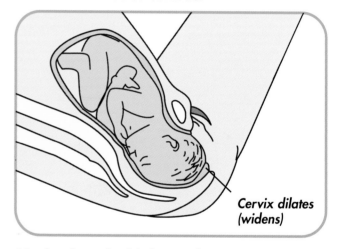

Moving into the birth canal

FACT FILE

Twins are two babies who share the same uterus during a single pregnancy.

They can be identical or non-identical, either of the same or opposite sex. Triplets are three babies and quadruplets, four!

How do twins occur

Twins can occur in two ways. Firstly, when just one egg is fertilised but within days it splits, each half developing into an identical baby. Secondly, when two eggs are released at the same time and are fertilised by separate sperm resulting in non-identical twins.

When are most babies born?

Through the months of the year, slightly more babies than normal are born during August, and slightly fewer in April. The most common time of day for births is in the middle of the night, between 3 and 4 am. The least common time is twelve hours later, around 3 pm in the afternoon.

What colour are a baby's eyes

Nearly all babies have grey–blue eyes at birth. This is because the coloured part of the eye, the iris, has not yet developed its full set of microscopic cells which produce the colouring substance, or pigment. The eventual colour develops over a few months.

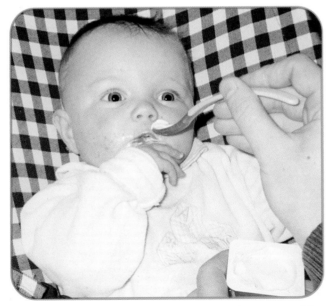

A blue-eyed baby stares at the bright world

What is a baby's first food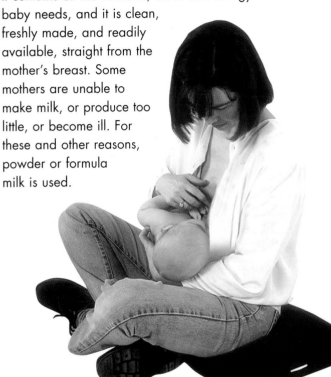

Mother's milk is the perfect food for new babies. It contains all the nutrients, fluids and energy that the baby needs, and it is clean, freshly made, and readily available, straight from the mother's breast. Some mothers are unable to make milk, or produce too little, or become ill. For these and other reasons, powder or formula milk is used.

A baby feeds

FACT FILE

For every 100 new baby girls, on average, there are 105 new baby boys.

At the time of birth, the womb (uterus) contains the most powerful muscles in the human body – 40 times their size before pregnancy.

BABIES AND TODDLERS

During the first years of life, a child develops rapidly and learns a great deal about social roles and behaviour. It is during this time that they learn to walk and talk. Young children are like little sponges, soaking up information about the world around them.

When does a baby first smile

Babies usually start smiling from the age of six weeks. Smiling is an important part of communication. A smile can indicate that a baby is happy and content, or wants attention of some kind. Long before the first smile, babies want to communicate with people. They communicate by copying the facial expressions of their parents, holding the gaze of others, widening their eyes or moving their tongue.

Can toddlers walk before they crawl

Most babies and toddlers go through the stages of development in the same order, but their age at each stage can differ greatly. A typical baby can roll over at nine to ten weeks of age, sit up unaided at eighteen to twenty weeks, stand up supported from six or seven months and crawl at eight months. He or she may walk unaided at eleven to twelve months. Some babies don't bother with crawling. They learn to roll sideways or bottom-shuffle, and then go straight to walking.

Sitting and standing are important stages, or 'milestones', of development

Why do toddlers like bland foods

The sense of taste helps to warn us of foods which are bad, rotten or poisonous – and these often taste sour or bitter. A toddler follows their in-built reactions, or instincts, and avoids these tastes.

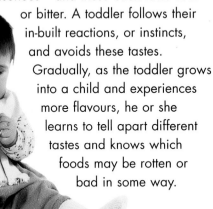

Gradually, as the toddler grows into a child and experiences more flavours, he or she learns to tell apart different tastes and knows which foods may be rotten or bad in some way.

Babies don't eat food – they wear it!

What does cutting teeth mean ?

Between seven and twelve months, teeth start to grow and push through the baby's mouth. They come from little tooth 'buds' which were formed while in the womb. Usually, the first to appear are the 'central' incisors, or front teeth, followed by the 'lateral' incisors. The baby is likely to drool more, eat and sleep less and be generally uncomfortable – which is not surprising!

Most babies have a full set of milk teeth at three years. By the time they are six, they start to lose them as their permanent teeth come through.

Is it true that newborn babies can swim ?

Having spent months suspended in fluid in their mother's womb, babies feel at home in the water. All babies are born with a swim and dive reflex. This means that they automatically hold their breath under water and are able to swim for short distances. This reflex is strongest when a baby is

under three months old. After this time babies can develop a fear of the water – but they can easily learn to swim properly when they are older.

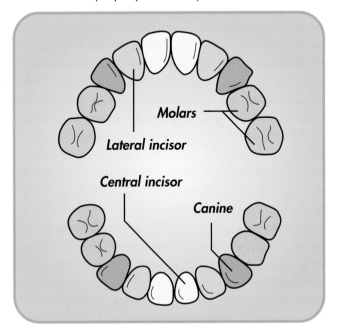

Molars

Lateral incisor

Central incisor

Canine

The positions of milk teeth

GROWING OLDER

Growing 'up' is not just getting taller, or even becoming bigger all over. It means a whole host of changes inside the body, especially during the teenage years. As many more years go by, things change again and bodily systems start to slow down.

What is puberty

Puberty is the time when the sex organs (reproductive parts) become active and start to function. This phase mainly involves physical growth and development.

Puberty is part of a longer period called adolescence, which involves many mental and emotional changes, changes in friendships and social life, and increasing independence from parents.

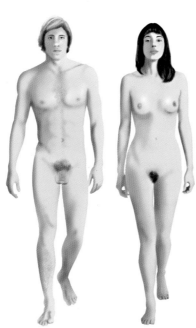

Adolescent boy and girl

What is the menstrual cycle

The menstrual cycle is something that is necessary for reproduction. During the menstrual cycle, the ovary releases an egg in an event called ovulation. If the egg is not fertilised it will die and be absorbed by the woman's body. After this, hormones cause the uterus to shed its lining in a process called menstruation.

Egg matures in ovary

Ovulation (egg released from ovary)

Egg breaks down

Womb lining is shed (menstruation)

Level of oestrogen

End of menstruation

Level of progesterone

| 1 day | 7 days | 14 days | 21 days | 28 days |

Start of menstruation

The menstrual cycle lasts on average 28 days

When do women stop having babies

The menstrual cycles of women usually stop around the age of 50–55 years. This stage of life is called the menopause, and it means that ripe eggs are no longer produced by the ovaries.

However, with modern medical techniques such as IVF (in vitro fertilisation or 'test tube babies') it has become possible for women to conceive at a much later age. Most men carry on producing sperm for longer, into their 60s or even 70s, so they can still be fathers.

Why do voices 'break'

The bodily changes of puberty involve extra growth of the voice box, or larynx, in the neck. The strip-like vocal folds (cords), which vibrate to make the sound of the voice, become thicker and stiffer. They vibrate more slowly, making the voice deeper. This happens most noticeably in boys, whose voices 'break' or 'crack' during puberty, but also to an extent in girls.

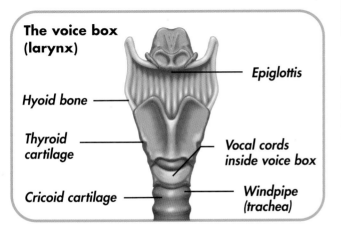

The voice box (larynx)

Epiglottis

Hyoid bone

Thyroid cartilage

Vocal cords inside voice box

Cricoid cartilage

Windpipe (trachea)

Why does skin become wrinkly

Skin contains tiny threads, or fibres, of the substance elastin. Like millions of microscopic elastic bands, these make the skin spring back after it is stretched. With age, the amount of elastin becomes less so the skin becomes stretched and loose, and puckers into creases and wrinkles.

FACT FILE

On average, in most developed countries, puberty happens two years earlier than it did 100 years ago.

This is probably due to better foods with healthier nutrients which allow faster growth.

EAT WELL, STAY FIT!

The body benefits from moderate and regular exercise, two or three times weekly, and also a wide range of food and drink. The body needs a well-balanced diet to provide it with the substances it requires for energy, growth and repair.

Which foods are healthy

The body can cope with most kinds of food. The main problem is too much of any one kind, whether it's crisps or carrots. A varied diet is best. Fresh fruits and vegetables are especially healthy. People benefit from four or five portions of them daily – a portion being a piece of fruit or a helping of vegetables.

Fresh fruits are very healthy

Which foods are not healthy

Too many fatty foods can be bad for the body, especially foods with lots of animal fats such as fatty meats or full-fat cheeses. In particular, they affect the heart and blood vessels. Also, too much food in total is unhealthy. Being overweight increases the risk of many problems such as heart disease, high blood pressure, breathlessness, diabetes and breathing disorders.

Fatty and fried foods can lead to serious heart problems

Why is smoking harmful

Smoking tobacco is a very bad idea. It is linked with diseases such as heart disease and cancers of the lungs, throat, gullet, bladder and many other parts. It can cause breathlessness, wheezing and increased risk of infections such as bronchitis and pneumonia.

Worldwide, there were about 100 million tobacco deaths during the twentieth century alone.

Which sport or exercise is best

The type of exercise or activity does not matter too much, provided it is done in moderation. We are more likely to continue with regular exercise if:

- we enjoy it and it's good fun
- it fits in with our daily or weekly routine
- it doesn't involve too much special equipment, travel or money.

It could be a team game like football or hockey, a sport such as tennis or squash, or swimming, cycling, jogging, horse riding ... the choices are endless!

How often should we exercise

Most people benefit from at least two exercise sessions each week. These should be enough to cause faster breathing and a more rapid pulse rate, which shows the lungs and heart are getting a workout. A bit of sweat is also usually a good sign.

Regular exercise benefits the body

Swimming is a great way to get fit

FACT FILE

Smoking while pregnant can harm the unborn baby.

Cigarette smoke contains over 4,000 chemicals, including over 50 known carcinogens (causes of cancer) and other poisons.

GLOSSARY

Alveoli

Tiny air sacs in the lungs that carry the oxygen from the air into the blood.

Canines

The taller, pointed teeth to the sides of the incisors at the front of the mouth, good for tearing tough food.

Capillary

Delicate, thin-walled blood vessels, some of which are finer than human hairs.

Carbohydrate

One of the main food groups, relatively high in sugar and starch – includes bread and potatoes.

Carpals

The eight small bones of the wrist which join the hand to the forearm bones.

Cartilage

A tough, relatively flexible tissue that can be found in the nose and ears and in joints – most of the skeleton is made from harder bone.

Cell

The smallest unit of an organism (living thing) that can function on its own.

Cerebellum

The part of the brain that coordinates movement and balance.

Cerumen

Ear wax that plays an important role in cleaning the ear canal.

Cilia

Tiny hairs on cells, such as those in the nose that, when touched by odorous particles, send messages to the brain for our sense of smell.

Digestion

The process by which food is broken down in the stomach and intestine into substances that can be absorbed into the body.

Enamel

The hard, white substance that covers each tooth.

Eustachian tube

A tube inside the ear which acts as a pressure gauge, letting air both in and out.

Fertilisation

The point at which the man's sperm joins with the woman's egg and the life of a baby starts.

Fibrin

A protein formed when blood clots, making a network that holds the red blood cells and platelets together.

Fibrous

Consisting of, or resembling, fibres, as in fibrous tissue.

Foetus

The tiny, growing baby in the womb, from the second month of pregnancy.

Fontanelle

A small gap in the top of a baby's skull which can appear as a soft spot.

Genes

Inherited, genes carry instructions for how the body grows, develops and carries out life processes.

Haemoglobin

A protein that gives the red colour to blood cells.

Hair follicle

A tiny cavity in the skin from which hair grows. It houses the only part of the hair which is alive.

Hormone

A chemical substance that carries a signal from one cell to another. Produced in the endocrine glands, it is transported in the blood.

Incisors

The chisel-edged teeth at the front of the mouth.

IVF

In Vitro Fertilisation – If a woman is unable to conceive naturally, egg cells can be removed from the woman's ovary and fertilised by the sperm outside the body, before being implanted in her womb (uterus) when pregnancy proceeds in the normal way.

GLOSSARY

Keratin

The substance (a protein) from which nails and hair are made.

Melanin

Black or dark brown pigments in the hair, skin and eyes of humans.

Metacarpals

The five bones that form the palm of the hand.

Myofibril

Muscle fibre that consists of actin and myosin proteins.

Nerve

A fibre in the body that transmits messages between the brain or spinal cord and other parts of the body.

Nutrient

Any substance that nourishes the body – repairing it when needed and keeping it in good working order.

Odorous particle

A 'smelly' airborne particle that floats into the air and into our nose, touching micro-hairs (cilia) which give us our sense of smell.

Oestrogen

A hormone which produces and maintains female characteristics.

Olfactory

Relating to the sense of smell.

Ossicles

The three tiny bones of the inner ear, known as the malleus (hammer), incus (anvil) and stapes (stirrup).

Paresthesia

The medical term given to 'pins and needles' – a loss of sensation in areas where the blood pressure has been affected.

Phalanges

The two bones of each thumb and three bones of each finger.

Pigment

A substance that is found in animal or plant tissue that produces the dominant colour.

Protein

Compounds that form part of the body tissues and an important part of our diet.

Plasma

The yellowish fluid part of the blood.

Platelet

A minute particle in the blood which helps it to clot.

Receptor

A nerve ending which responds to touch and sends messages to the brain, for example, touching a hot surface with our fingertips and then moving them away quickly.

Retina

The light-sensitive lining at the back of the eyeball.

Sebaceous gland

Small glands in the skin that secrete sebum into the hair follicles and over the surface of the body – except the soles of the feet and the palms of the hands.

Sound waves

Waves that enable sound to travel, as in the sound travelling from the mouth through the air to our ears.

Synovial fluid

A clear, lubricating fluid secreted by the membrane lining the joints, making joint movement smooth.

Testosterone

A very strong hormone produced by the testes in the male.

Umbilical cord

The long, tube-shaped cord joining the foetus to the placenta, along which vital nutrients are passed from the mother and waste is passed back.

UV rays

Ultra-violet rays from the Sun which can cause wrinkles and premature aging as well as sunburn.

Vena cava

The two large veins that carry oxygen-depleted blood to the heart.

INDEX

Picture credits

Main illustrations by Meme Design and Stephen Sweet.
Additional Photography and illustrations by Beehive Illustrations,
Sushumu Nishinaja and the Science Photo Library.

Collect them all!

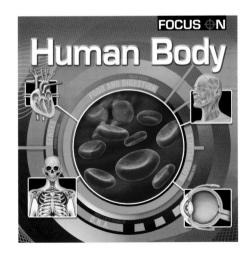